Rocks and Minerals for Little Eyes

By Patrick Nurre

The Northwest Treasures Curriculum Project
Building Faith for a Lifetime of Faith

Rocks and Minerals for Little Eyes

By Patrick Nurre

Rocks and Minerals for Little Eyes
Published by Northwest Treasures
Bothell, Washington
425-488-6848
NorthwestRockAndFossil.com
northwestexpedition@msn.com

All photographs and images are by Vicki Nurre or Patrick Nurre. Rock samples are from the author's private collection.

Rocks and Minerals for Little Eyes

Introduction for Parents

Dear Parents, people who believe the Bible are ridiculed when it comes to things like the Creation and the Genesis Flood. I have seen many Christians compromise their trust in the Bible because of what people say. While we wrestle with our faith at times, think of what your child will go through when he/she is older. We must start now to teach them about how to view the earth from beginning to end. The Bible is more than a philosophy for life. Much of the Bible is recorded history supervised by the One who was there and saw it all. Teaching our kids to study the rocks in light of what God has spoken, will give them a solid foundation for trusting God's word in the future as they are challenged again and again. As you progress through the different learning levels of your kids, you can build off of the material presented here. Kids are naturally drawn to rocks. Let's build off of that interest by telling them the whole story – not just the Sunday morning portion of it.
There are a few things that a child of this age should be made aware of and that will serve as the foundation for the development of a Biblical view of Earth history.

It is assumed that most children will not be able to read this text themselves, and as such, is designed for the parent to read to the child, in a one-to-one dialogue, asking the questions and probing for understanding. The activities are also designed for the parent to take the lead. You may feel free to cover this material at your and your child's pace of learning.

The questions and activities appear in red in the text. I would suggest that you help your child keep an activity notebook of their answers to questions, and what they have discovered in the activities. Depending on their age, you may have to do the writing. But they can dictate what they learn. You simply become their voice on paper. They can also keep notes using pictures that they draw. The goal is to teach our kids to practice observing and recording, which are very important scientific skills.

God bless you in your endeavor to teach your children the true and real history of the earth.

Rocks and Minerals for Little Eyes

Lesson I: The Creation of Earth

The word geology comes from two Greek words – *geo*, meaning earth, and *logi*, meaning the study of. So, geology is the study of the earth. When we study the earth, we should be curious about who made the earth, how and why it was made, all the rocks we see and the future of the earth. Aren't you a little curious about these things?

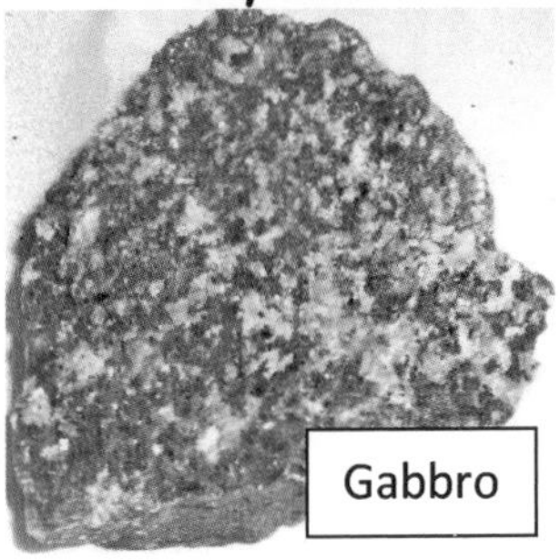

Gabbro

But, in order to answer these questions, we need to know someone who was actually there when the earth was made. Let's see. Who could that be?

Think back to your favorite birthday party? Were you there? Who else was there? What kinds of things did you do? ***Make a list of the things you remember.***

The things you listed are called *evidences*. These are additional things that help me have trust in what you say.

Now, I wasn't there when you had that fun birthday party. How will I know about your birthday party? Well, you told me so. I believe you because you told me. And, I can also believe you because of the *evidences* you listed above. The same would be true of your parent's marriage. You know all about their marriage because they told you and because of the pictures and the reports from others who saw their marriage. Were you there? No, but others were. And so, you believe them.

The same is true when the earth was made. You weren't there. Your parents weren't there. Your friends weren't there. Your teachers weren't there. But God was there. And He has told us so in His Bible.

Memorization: ***Genesis 1:1***

So, God was there when the earth was made and He has told me so in His Bible. Here is a great little saying that I learned early on in my Christian life: *"God said it. I believe it. That settles it."* This is what faith is all about. Faith is trusting the word of the One who was actually there when the earth was made.

But just like you had evidence of your birthday party, so God has left evidence that He made the earth. ***Can you think of evidence that God has left us that He made the earth?***

Now that we have begun a good habit of listening to what God said about how the earth was made, do you think God has anything to say about the rocks that we see around us?

Of course God's main interest is in people, isn't it? How do we know this? Well, God's Bible tells me so.

Memorization: ***John 3:16***

OK. So, where did all the rocks come from? Remember, rocks are part of the evidence that God left us about what He did when He made the earth. Some rocks are pretty and they show the work of God Who is a Master Artist. Some are useful to us and therefore show God's care for us. Some rocks are just interesting, like God is. I think rocks from volcanoes are especially interesting. ***What are some of your favorite rocks?***

So, some rocks show something about our God as our Maker. But some rocks, although interesting to look at, seem to tell a different story. Some rocks are twisted and crushed. ***Can you find a rock that looks like this? Can you think of a story from God's Bible that might tell us how these rocks got this way?***

If you guessed the story of Noah's Flood, you would be correct. Let's get God's Bible and read about Noah and his life from *Genesis chapters 6-8*. Now, remember, God was there when the earth was flooded. So, we can trust what He says about this event.

When you have finished listening to the story, list all the phrases that might tell us something about what happened to the earth during Noah's life.

So most of the rocks we see around us will definitely tell us something about God, our Maker. But most of the rocks we see around us will tell the story of how God destroyed much of the earth He had made. ***Why did He do this?***

Do Activity #1.

Cross bedding, Utah

Lesson II: The Flood

Memorization: ***Genesis 6:5*** To me, this is one of the saddest verses of God's Bible. God made a beautiful Earth, but because of the evil of man, God had to destroy it. And much of the earth we see today is evidence of God's judgment. The rocks today tell us that God is watching man and will hold him responsible for what he does. Even most of the dinosaurs died as a result of man's evil. How do we know this? Because the fossils are evidence of this.

Fossil fish

Fossils are the remains of once living plants and animals preserved in the rocks. Have you ever found a fossil? Fossils tell us how powerful the Flood of Genesis was. This was no ordinary flood. This Flood had never happened before and it has not happened since. Most fossils I have found tell me a story that matches what I read in God's Bible.

Memorization: ***Genesis 7:23***

Another piece of evidence about the Flood that we can see today is found in all the volcanoes that are around the earth. Some volcanoes have stopped erupting and only their mountains are left. But some are still erupting. Where did all the volcanoes come from? I think I know. Let's see if you can guess. Read this verse from God's Bible.

Memorization: ***Genesis 7:11***

So, it just didn't rain for forty days and nights, but the whole Earth was torn apart. Huge earthquakes must have taken place. Many mountains must have crumbled. Tremendous amounts of magma (lava) must have shot up from below the earth to form volcanoes. ***Can you think of other things that must have happened with so much earth movement?***

All around the earth we can still see what has been left of the Flood. The earth is weak in many places. Earthquakes, tsunamis, tornados, volcanic eruptions are all a part of the weakened Earth as a result of the Flood. God helps us when these things affect us, but He has left them as evidences of the tremendous Flood that once destroyed the earth.

This is Biblical Geology! The rocks all around us tell a story of destruction and awesome power, like what God's Bible tells me. Some people may want to tell a different story, but they weren't there, were they? Who would you rather trust? God, Who was there, or the people who were not there, but tell a different story?

Do Activity #2.

Lesson III: What is the Earth Made of?

Why talk about this in a rocks and minerals class? The phrase, "In the beginning, God..." tells us what took place at the beginning of time and Earth and why. The center of God's focus in this first chapter of Genesis was creating and forming a very special planet for man - Earth. And therefore we want to know what the Earth was made of in the beginning. Even the creation of the sun, moon and stars, on Day four of Creation week, were to help fulfill this purpose. The sun, moon and stars were to give light on the Earth and to mark seasons, days and years. This would only be important if it was for the sake of man, however. There is no other explanation for "seasons, days, months and years." Think about it: do animals count seasons, days, months and years? NO! Man does!

Quartz chrysoprase

Secular scientists always start their history of the Earth with, *in the beginning, 15 billion years ago, the Big Bang, etc.* As God made a very special place in all of His creation, called Earth, it is important that we know a little about the composition of the Earth. What is it made of?

Geologists know quite a bit about the *crust* of the Earth. But no one has ever seen the inside of the Earth. Geologists think the Earth has a *crust*, a *mantle* and a *core*. Geologists have not drilled all the way through the crust of the Earth, so they don't know for sure what is below the crust! Geologists just guess at what might be there. They use sound waves to try and determine what is there. They also look at certain rocks that are different from those found in the crust. Some rocks are heavier than what is found in the crust. And this gives them the idea that perhaps the inside of the Earth is made of heavier rocks. That might be, but we don't know for sure. ***Take out the sample of a heavy rock and a light rock from your kit. What do you notice?***

Geologists think the Earth may look like this:

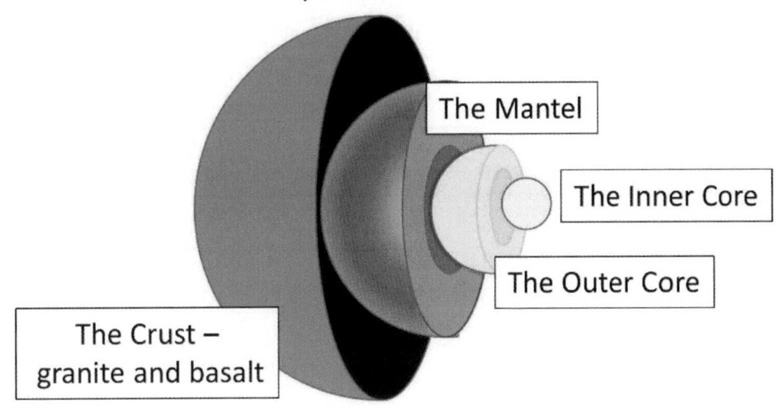

Do Activity #3.

Pegmatite

Lesson IV: Minerals, Elements and the Crust of the Earth

God used the atom as the basic building block for all created things. The atom is made up of charged electrical particles called protons (+charge), electrons (- charge) and neutrons (neutral charge). This all means that the atom is a balanced created thing that has been designed to be that way so that God's universe runs smoothly. The atom may look like this, but we don't know for sure. It is so small! *Five million hydrogen atoms would fit on the head of a pin!*

Quartz crystal

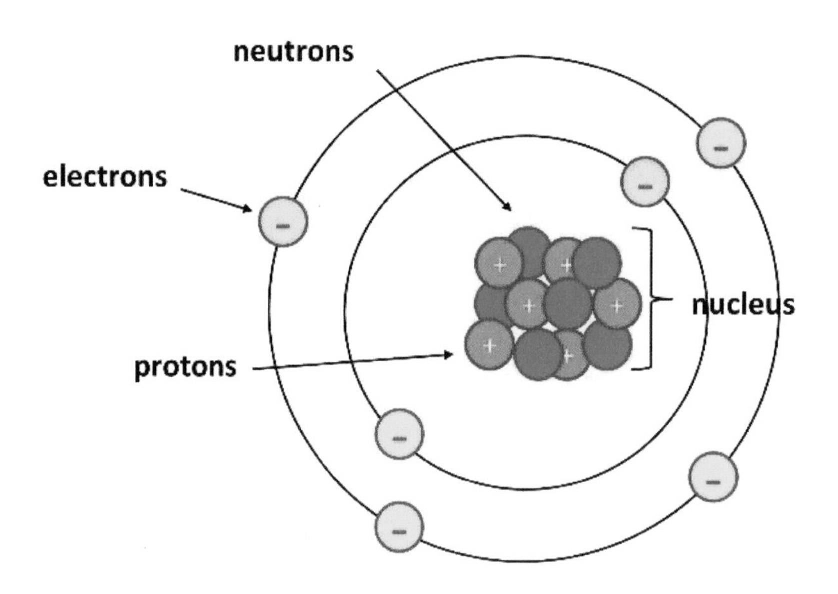

Minerals are made up of **elements** that are the basic atoms of all things. Each element is different and has its own set of charged particles arranged in a very special way. For example, let's just take a look at the atoms that form the simple **mineral** called quartz. It is formed from two different **elements** called silicon and oxygen.

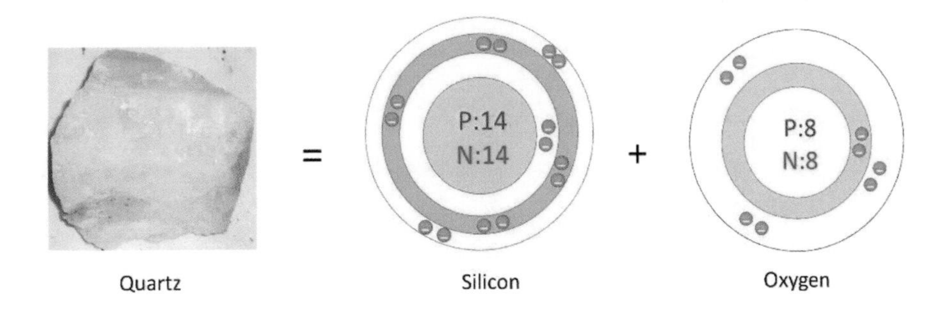

Quartz is formed from silicon and oxygen.

The element **silicon** has 14 protons, 14 neutrons and 14 electrons. The element **oxygen** has 8 protons, 8 neutrons and 8 electrons. Combined in a very special way, these two elements give us quartz, a mineral that is found in almost every rock.

Every planet and star is made up of the elements that God created. Secular geologists think that because hydrogen and other elements exist on other planets that all the planets are related. In other words they think that all the planets and stars came from the same ball of stuff in the beginning 15 billion years ago. But the Bible tells a different story. The elements are common to other planets because they have a common maker - God. He used the same elements and patterns all throughout His creation. But only the Earth has been put together in a very special way different from every other planet. It is the only planet that we know of that has been prepared for life. The Earth's **crust** is made up mostly of just 8 **elements** that have been combined in very special ways to form most of the **rocks** we have on Earth. The *"other elements"* include elements like carbon, fluorine, titanium and hydrogen, among many others.

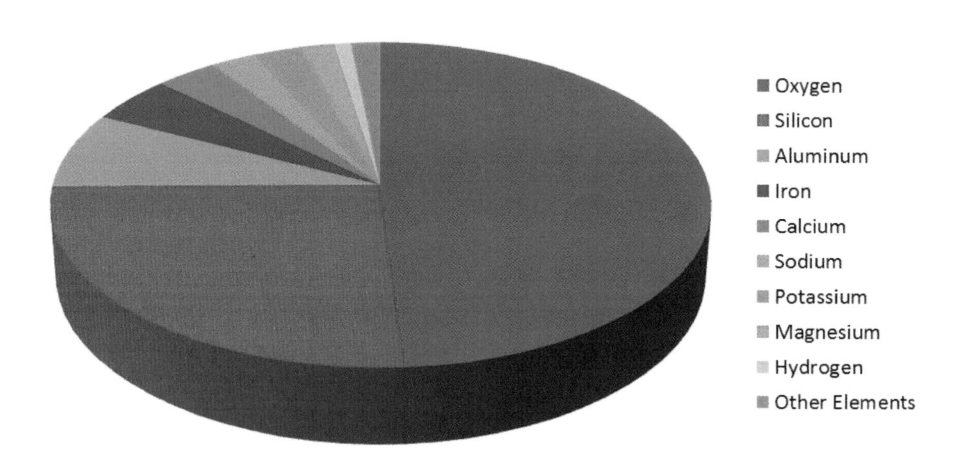

Notice also that all these elements are necessary for life? You get all these elements in the foods that you eat which come from the basic created stuff that God put here 6,000 years ago.

These **8 elements** have also been combined in very special ways to form a basic set of **rock-forming minerals**. We call them "rock-forming", because they form most of the rocks on the Earth. We will talk more about these later.

Do Activity #4

Calcite crystal on limestone

Lesson V: Rocks are Made of Minerals

You remember that minerals are made of elements. Geologists also know this: **Rocks are made of minerals.** When two or more **minerals** are combined in very special ways, they give us the beautiful rocks we see all around us.

Granite is used for lots of cabinet tops. It is beautiful. You have these minerals and the rock called granite in your kit. *Take them out and look at them closely. Describe what you see.*

The Rock-forming Minerals

<u>The Dark Minerals</u>

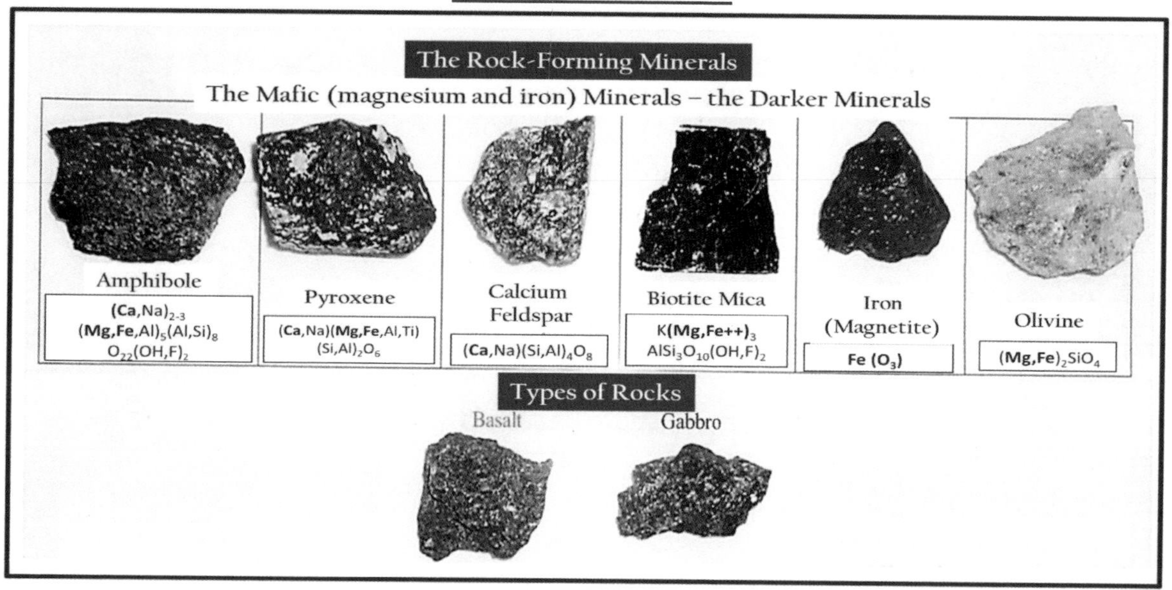

The Light Minerals

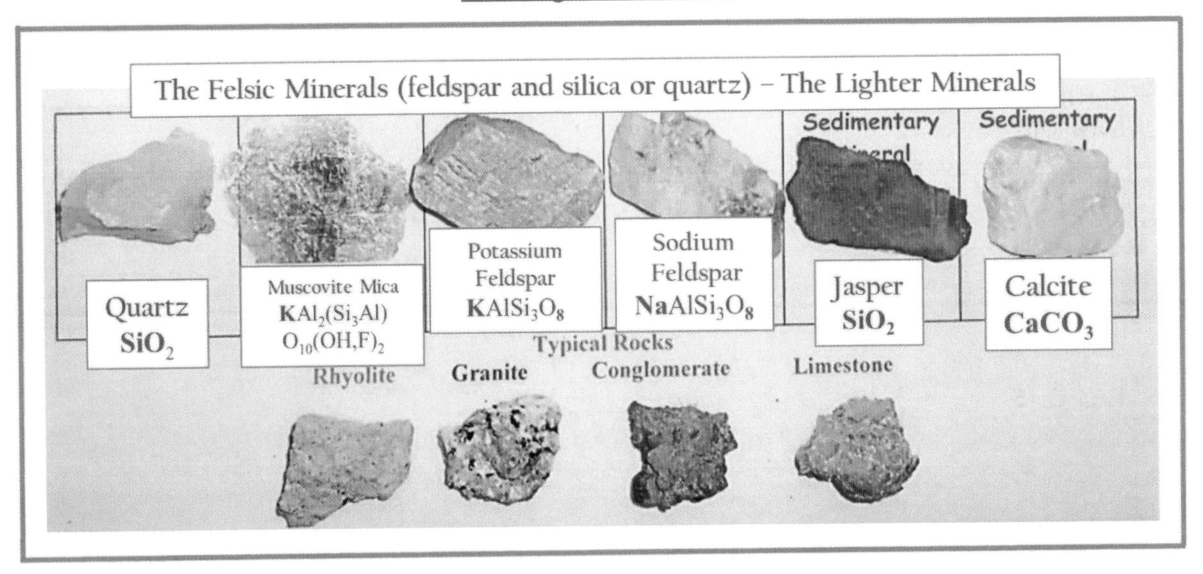

Notice that all the above **minerals** are combinations of **elements** arranged in a very special way. The combinations are listed under each mineral. These elements form the crust of the Earth and the rock-forming minerals. See if you can unscramble the symbols under the names of the rock-forming minerals and name the elements. The list of elements is listed below to help you.

O- Oxygen
Si = Silicon
Al = Aluminum
Fe = Iron
Ca = calcium
Na = sodium
K = potassium
Mg = magnesium
H = hydrogen
Ti = titanium
C = carbon
F = fluorine

All the **rock-forming minerals** are made of the 8 most basic elements in the Earth's crust plus a few more we call *incidental* because they don't occur in all minerals. All the elements in the Earth's crust are combined in very special ways to form these **rock-forming minerals**. In the list of the minerals above, which elements are the most basic elements of the Earth's crust? They are the ones in bold letters. *Now, here is a little secret - if you can learn these minerals in the above chart, you will be able to identify just about every rock you can find!*

Now, here is the most perplexing question for geologists - how were the **elements, minerals** and **rocks** made? The Bible clearly teaches us that God is the Maker of all things. Modern geologists don't accept this and think they know better and so they guess. But no geologist has ever seen granite or the metamorphic rocks form. The only rocks they have seen forming are some of the volcanic rocks. The Book of Genesis reveals to us that God made all things. That is the most simple and easy answer to that perplexing question that geologists just scratch their heads over.

Do Activities #5 and/or #6.

Calcium feldspar

Lesson VI: What are the Differences Between the Rocks?

Once you start collecting rocks, you will notice some basic differences among them. Here is a list of some of those basic differences.

- Some rocks you find will be **coarse-grained** rocks; rocks in which you can see the mineral grains. *Examples are:*

Gabbro and Granite

- Some rocks you find will be **fine-grained** rocks; rocks in which you cannot see the mineral grains. *Examples are:*

Basalt and rhyolite

- Some rocks you find will look like they have been deformed in some way. *Examples are:*

Gneiss, phyllite and slate

- Some rocks you find will look like they have been stacked like pancakes. Some of these will be very fragile. *Examples are:*

Shale and sandstone

- All these rocks can also be grouped according to whether they are **light colored** or **dark colored**. Generally, the **minerals** pictured earlier form all the light and dark colored rocks. Take a look at the basalt and rhyolite in the above pictures. One is dark colored - the basalt; and one is light colored - the rhyolite. That is because the **dark colored minerals** form the basalt and the **light colored minerals** form the rhyolite. *Look at the chart of rock-forming minerals above and say the name of each mineral and whether it is dark or light.* This will be another way you can identify the rocks all around you.

- The different variations in the colors in the rocks are all due to the addition of other different **elements**. For example, iron causes things to turn brown or orange. Copper causes things to turn brassy or even green and blue.

Do Activities 7, 8 and/or 9.

Lesson VII: The Types of Rocks - Plutonic

Geologists organize rocks according to how they ***think they formed***.

Now, that is interesting because many of the rock types geologists have never seen forming. So, how do they know such and such a rock formed the way they think? They don't. They guess based on a view called **uniformitarianism.** This big word means that the Earth has been growing all by itself from the very beginning of its origin, 4.6 billion years ago, according to geologic processes they see going on today. Well, what about the Creation and the Flood spoken of in the Bible? Modern geologists reject these two events and therefore they believe that the Creation and the Flood are not important to the study of the Earth. That is a big mistake, because the Genesis Flood was a real historical event that had global consequences.

So, geologists believe that there are 3 rock types - *igneous rocks*, meaning fire-formed over hundreds of millions of years of heating and cooling, *metamorphic rocks*, those rocks changed by intense heat and pressure over hundreds of millions of years and *sedimentary rocks*, those rocks formed through the slow accumulation of sediments mixed with water, mud and living things. Modern geologists believe that these rocks were formed over millions of years. But most of these rocks geologists have never seen forming. They have never seen granite form. They have never seen the metamorphic rocks form. And they have never seen the formation of most of the sedimentary rocks. What about volcanic rocks? Geologists organize these rocks under igneous rocks because they have seen at least some of them form out of fire.

A Different View

Let's take a different approach - one based on the historical Biblical account. According to what is recorded in the Book of Genesis, I would divide the rocks into **four groups** this way. Let's look at the first group:

Plutonic rocks - those rocks that were formed at the beginning of creation. Modern geologists use the word *plutonic* to describe these rocks. That word comes from mythology, the god of the underworld, Pluto. Geologists call rocks like granite, plutonic because these form the foundation of the Earth and are found deep within the crust of the Earth. Some were formed during the Flood, but most were most likely formed at the beginning of the Creation. These would include the coarse-grained rocks. ***Take a look at the coarse grained rocks (plutonic) in your kit. Describe what you see.***

Do Activity #10.

Granite, Yosemite National Park

Lesson VIII: The Types of Rocks - Volcanic

The second group we will look at is volcanic rocks – those rocks that form when lava is pushed out of and onto the Earth. Geologists get this name from the Island of Volcano, which is a volcano. The word volcano comes from the god of fire, Vulcan, in mythology. Geologists have called volcanoes "the crucible of creation". They believe that volcanoes are responsible for bringing about our water, soils and the beginnings of life. However, volcanoes are destructive to man, not creative! I don't believe they were part of the beginning of the Earth. They are rather cracks and weaknesses or holes in the Earth caused by the great Genesis Flood. These would include the fine-grained rocks. ***Take a look at the fine-grained rocks (volcanic) in your kit. Describe what you see.***

Obsidian

Do Activities #11, #12 and/or #13

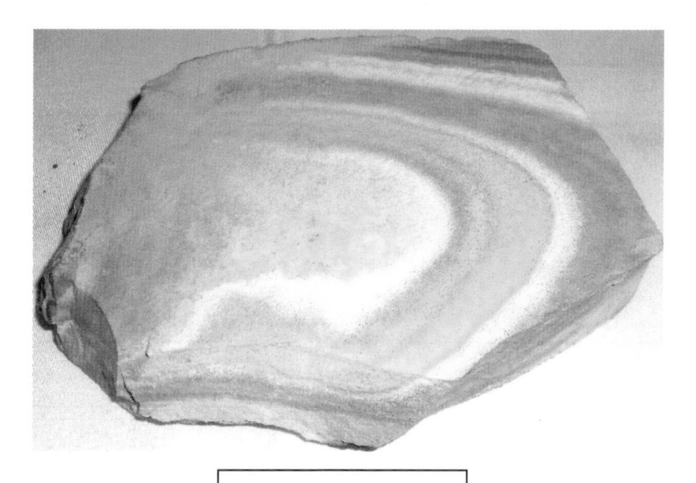

Rhyolite lava

Lesson IX: The Types of Rocks - Metamorphic

The third group we will look at is Metamorphic rocks - those rocks that were changed by the movement of other rocks against one another during the breaking up of the fountains of the great deep (Genesis 7:11) and afterward as the mountains rose and valleys sank down during the Flood (Psalm 104:5-9). These would include the changed rocks. They were changed because of the heat that came from rubbing blocks of rock against one another. This would have happened rapidly, not over hundreds of millions of years of slow movement. The Genesis Flood lasted for about 371 days. ***Take out the changed rocks (metamorphic rocks) from your kit and look at them. Describe what you see.***

Do Activity #14.

Schist

Lesson X: The Types of Rocks - Sedimentary

The last group we will look at is the sedimentary rocks - those rocks that were laid down in or with water and mud - sediments. As the waters from the Genesis Flood would have carried millions of tons of mud from one location to another. These sediments would have been laid down in row after row of layers that would have eventually hardened, but hardened rather rapidly, not over hundreds of millions of years.

Sedimentary layers, Utah

Many of these sedimentary rocks have fossils in them. ***Take out the sedimentary rocks from your kit and look at them. Describe what you see. Do any of them have fossils in them?***

Do Activities #15, #16 and/or #17

And there you have it - your first course in rocks and minerals. I hope you will continue your study to become an expert at locating and identifying all the many different rocks around you. If you study them closely, you will begin to see the Biblical story they tell.

Do Activity #18

Sedimentary layers, sandstone, Utah

Lesson XI: Now, get out in the field!

Do Activity #19

Obsidian, Glass Mountain, California

Activities

*For all activities, keep an **activity notebook** to make appropriate comments, pictures or notes as needed. Some observations may need to be notated by an adult. Simply ask your child what they observe, and write what they say.*

Activity #1 - Density Column

This project is to help you to understand how watery sediments layer. This should give some insight into what happened during and after the Flood.

Materials:

Honey
Corn syrup
Molasses
Milk
Oil
Dish soap
Pancake syrup
Any other liquids you like
Tray
A glass jar (holds 1-2 C.)
Small pitcher
A few small items, like marbles, paper clip, eraser, piece of rock, sand, etc.

Procedure:

You are going to pour the liquids in, one at a time, into the jar. They are going to settle to different levels. Can you predict which ones will settle on the bottom? Middle? Top? Go ahead, now, and pour each liquid into the jar, one at a time. Watch carefully for movement of the liquids. A great way to add to this project is to add items and predict and observe whether the items sink or float and into which layers. Record your predictions about which items will settle on the bottom, middle, top. What predictions might you make about how sediments would have settled during/after the Great Flood? Record what thoughts you have about how sediments (and plants/animals) might have settled during and after the Flood,

Activity #2 – Splitting Rocks…with Water

In addition to volcanoes and great flooding, ice played a large role in sculpting and shaping our world in the aftermath of the flood. This experiment hi-lights the power of ice to change things.

Materials:
Balloon
Plastic container or aluminum pie plate
Plaster of Paris

Procedure:
Fill a balloon with a small amount of water and then cover with plaster. Allow it to dry. Put it all in a container and put in the freezer. The ice will break the plaster apart.

The freeze/thaw cycle causes rocks to crumble and break down into little rocks. When water gets into cracks in the rocks, this water expands during the freeze cycle, making the cracks bigger. Then when cracks fill up with water in the thaw period, this allows water to go deeper into the rock. Each time the water freezes, the crack gets deeper, and in time, the rock will split. The power of frozen water expanding can also be seen when you leave a glass bottle filled with liquid or an unopened soda can in the freezer. (If you have a spare bottle or soda, you might try this!)
Record your observations.

Activity #3 - Earth Model

Materials:
4 colors of modeling clay

Procedure:
Make a model of the earth's layers, using the picture in the text as a model.

Pahoehoe lava, Hawaii

Activity #4 - "There's iron in that thar cereal!"

You can find minerals in all kinds of things. In this activity, we will discover iron is in even some of our favorite cereals.

Materials:
Three or more boxes of cereal, including Total Whole Grain
Very strong magnet
Rolling pin or blender
Gallon zip lock bags
Hot water
Sharpie marker
Tape (optional)

Procedure:

1. Take your marker and label your zip lock bags with the cereal names.

2. Pour 3 cups of each cereal into the correct zip lock bag.

3. Use the rolling pin (or blender) to crush the cereal in the bags into as fine a powder as you can get. *This is important.*

4. If holes were created in the bags during crushing, transfer the crushed cereal to a new bag. Cover the crushed cereal with hot water.

5. Zip the bags shut leaving an air pocket in each bag. You can use a little duct tape to guard against leakage.

6. Allow the bags to sit for 30-60 minutes. Flip the bags over every ten minutes and give it a slight shake to mix your concoction.

7. The adult may need to help do these next steps because of the co-ordination required. Begin with the Total cereal. Take out your (very strong) magnet and hold it firmly underneath on the bottom of the bag (touching the zip lock plastic). You might consider using tape to help it stay attached. Start a slow swirling motion of the cereal in the bag. Be very careful to never let the magnet separate from the zip lock plastic surface or the iron will fall back into the mix. Keep the swirling going for several minutes to get the most iron. The iron needs time to sink to the bottom of the bag.

8. While keeping the magnet to the bags surface, flip the bag over so now the magnet is on top. You will see the tiny iron filings collecting around the magnet. These are real little bits of actual metal. The cereal company adds actual iron/metal because iron in other forms actually reduces the shelf life of the cereal. The acid in your stomach is strong enough to dissolve these tiny bits
of metal and release the elemental iron into your system.

9. Remove iron from bags by using the magnet and scissors to cut plastic around the filings or dump out into a bowl and submerge the magnet to attract filings. If you have a kitchen scale, weigh the filings from each cereal bag.

10. Repeat the same process with other cereals.

Be sure to record your findings in your notebook.

Activity #5 - Making Crystals (Epsom Salt)

Growing crystals doesn't have to take lots of time. Here are two crystal making projects that don't require lots of time and are so easy!

Materials:

Small bowl
½ C. Epsom salt
½ C. hot water (NOT boiling)
Food coloring (optional)

Procedure:

1. Place ingredients in bowl.
2. Stir until salt is dissolved. Don't worry if there are still crystals in the bottom of the bowl.
3. Place the bowl in the refrigerator undisturbed for 3 hours.
4. You can scoop the crystals out if you want to examine them more closely.

Graphic granite

Activity #6 - Making Crystals (Sugar)

Materials:

Glass jar with a lid
Sugar
Food coloring
Tape or pencil (any small stick)
Piece of twine
(optional) weight of some sort - paper clip, washer, etc.

Procedure:

1. Pour about 1 cup of granulated sugar into the jar, and mix 1 cup of water into it. Drop in 4 to 5 drops of food coloring, whatever color you like. Stir the mixture well until all of the ingredients are combined completely. Then, screw the cap onto the jar tightly and sit your jar on a countertop where it will not be disturbed.

2. Tape the twine the length of your jar to the underside of the jar lid, and screw lid onto jar. (Alternatively, tie string around a pencil and place pencil on top of jar with string hanging inside of the jar.) Make sure that the twine is hanging down far enough to touch the bottom of the jar. Your twine can be any weight, but a thicker weight twine will yield thicker crystals. (Attach a weight to bottom of the twine if desired.)

3. Keep an eye on your jar. Over the next few weeks you will start to see colorful crystals forming on the twine. You can take pictures periodically of the crystal growth and make a chart showing the rate at which your crystals are growing. After the crystals start to form, you can even add a few drops of food coloring of a different color and watch crystals of that color form on top of the first color.

4. Draw a picture in your notebook of your crystals.

Activity #7 - Rock Investigation

When scientists begin to investigate something, they have to make careful observations. We are going to observe characteristics about our rocks to get some practice in this.

Materials:
The rocks in your kit and any others that you want to add to them

Procedure:

Organize your rocks according to:
1. Size: from smallest to largest
2. Color: Lightest to darkest
3. Grain: Make two groupings: Is your rock smooth grain (showing one basic color) or coarse grain (you can see the minerals contained in the rock)?
4. Weight: lightest to heaviest

Basaltic glass

Activity #8 - Fracture or Cleavage?

Children should wear safety goggles during this experiment.

The shape of some minerals may be a clue to their identity. A mineral that breaks into to a geometric shape has good cleavage. Galena is a mineral with perfect cubic cleavage. If you hit a sample with a rock hammer, it will break into smaller cubes with each strike. Minerals that do not break into geometric shapes when struck will fracture instead. Quartz, the most common mineral, will fracture.

Materials:
Two old towels
Feldspar
Calcite
Galena
Hammer

Procedure:

To test a hard sample, sit a towel on the ground or on a hard surface. Place the sample on the towel and cover the sample with a second towel. Hit the sample with a hammer and examine the sample pieces for a geometric shape. Which pieces had a good geometric shape when broken? Can you draw those in your notebook?

Calcite

Activity #9 - Foaming Rocks

The calcite mineral in some rocks reacts when an acid touches the sample. Geologists use hydrochloric acid to test rocks. Children can test rocks for calcite using a weaker household acid such as vinegar or lemon juice.

Materials:
Limestone
Other varieties of rocks
Vinegar or lemon juice
Magnifying glass

Procedure:

Provide them with a hand magnifying glass to examine the rocks since the reaction will be on a smaller scale. Drop the lemon juice or vinegar directly on the rock sample to test it. If bubbles appear, the rock contains calcite. Common sedimentary rocks that are popular for this test are coquina and limestone, which can both contain calcite.

Make a chart of the rocks you tested, and whether or not bubbles appeared when testing them.

Limestone chalk

Activity #10 - Cookie Exploration

In this activity, you will be removing the chips, nuts, and/or fruit pieces from the cookie, giving the child an idea of the make-up of plutonic rocks. Plutonic rocks have minerals in them that can be seen with the naked eye. The chips/nuts/fruit represent the minerals.

Materials:

Chunky cookie with a variety of chips, nuts and/or fruit pieces in it
Toothpicks

Procedure:

Using a toothpick, carefully remove the various chips, fruits and/or nuts from the cookie. Talk about the makeup of granite and gabbro, for instance, and compare the cookie with these rocks in your kit.

Granite

Activity #11 - Intrusive or Extrusive?

Geologists classify volcanic rocks as either intrusive or extrusive. Intrusive igneous rocks form below the ground surface when the magma intrudes, or squeezes, into cracks or openings in the rock extending from the magma chamber. Extrusive rocks form from lava after it expels from an erupting volcano. An igneous rock's texture is often the key to determining if it is intrusive or extrusive. Extrusive (volcanic) igneous rocks have small to no visible mineral crystals. The intrusive (plutonic) rocks have medium to large mineral grains.

Materials:
Granite
Pegmatite
Obsidian
Basalt
Magnifying glass
Any other rock samples you like

Procedure:

Give your child several rock samples such as granite, pegmatite, obsidian and basalt. Let her examine them to identify whether they are intrusive or extrusive rocks. Make a chart in your notebook of which rocks were extrusive and which were intrusive.

Activity #12 - Floating Rock

Pumice and scoria are two volcanic rocks that look very similar. Children enjoy testing the difference between these two rocks because one sample will float and one sample will sink. Pumice floats because of the numerous air pockets in the rock. These are called gas pockets because initially, the pumice was filled with volcanic gas. This made the rock less dense, and consequently, lighter in weight.

Materials:
Pumice
Scoria

Procedure:

Fill a glass with water and give your child one sample of pumice and one sample of scoria. Place each sample in the water. The sample that floats is pumice and the sample that sinks is scoria.

Pumice

Activity #13 - Lava Rock Candy Dessert

In volcanic rock the rocks and minerals are fully melted into lava - this gives them a rather uniform texture and composition. Although no one has ever seen metamorphic rocks form, it is thought that metamorphic rocks form when sedimentary rocks have been subjected to heat and pressure and can be composed of many types of rocks at once so they often have a variety of shapes and textures within the same rock. In the following activity, you will use chocolate to make rock models that they will certainly want to examine more closely.

Lava Rock Chocolate

Materials:
Different types of baking chips - (white chocolate, peanut butter, butterscotch and/or chocolate)
Microwave safe bowl
Wax paper
Some type of mold to cool the chocolate in (can be a bowl, muffin pan, etc.

Procedure:

1. Grease or line your rock mold with wax paper.

2. Explain to your children that the different types of chips are different types of minerals and the microwave is a volcano that is going to melt them into lava. We are now going to make a dessert that is like a lava rock.

3. Let your children pour each kind of chip into a microwave safe bowl.

4. Put it in the microwave and heat until melted (about 2 minutes, though you should check it out every 30 seconds to stir it.)

5. When fully melted, stir it all together so that it all looks uniform, and pour it into the wax paper lined mold.

6. Put it into the refrigerator to cool.

Describe in your notebook what you see. Enjoy your rock chocolate!

Sheepeater Cliffs, Yellowstone National Park, basalt columns

Activity #14 - Making Metamorphic Rock Candy

Metamorphic rocks are thought to form through heat and pressure. This activity simulates that process.

Materials:
Graham crackers
Chocolate bar
Marshmallows
Gummy bears/worms or similar candy
Paper plates
Large flat-bottomed pan or bowl
Cooking spray

Procedure:

Layer the ingredients on a paper plate. Cook in microwave. Watch carefully as this may take as little as 30 seconds. Take out of the microwave. Spray second paper plate on underneath side. Place this side against the mixture. Place the flat-bottomed pan on top of that plate and press down. Remove the pan and second plate. Describe what you see. Record your observations. Eat, if you dare!

The Carpenter's Shop, metamorphosed sandstone, Israel

Activity #15 - Making Sedimentary Rock Treats

In this activity, the cereal represents the clasts in the rock, and the marshmallow represents the cementing agent. The main difference here is that sedimentary rocks usually do not involve heat. (The cementing agent in sedimentary rocks is iron oxide, calcite or quartz in a supersaturated solution. Heating the solution would alter the sedimentary rocks, and this is a process known as hydro-thermal- hot water -alteration.) We use marshmallow mixture for our project, since it closely mimics the saturated liquid. The cookie that you end up with appears much like limestone. Describe in your notebook what you see. Eat your cookies!

Materials:
3 T. butter
1 package (10 oz.) marshmallows
6 C. Rice Krispies or similar cereal
Various candies or cake decorating materials

Procedure:

1. **In large saucepan melt butter over low heat. Add marshmallows and stir until completely melted. Remove from heat.**
2. **Add cereal and other candies. Stir until well coated.**
3. **Using buttered spatula or wax paper evenly press mixture into 13 x 9 x 2-inch pan coated with cooking spray.**
4. **Cool. Cut into 2-inch squares.**

MICROWAVE DIRECTIONS:

In a microwave-safe bowl, heat butter and marshmallows on HIGH for 3 minutes, stirring after 2 minutes. *Microwave cooking times may vary.* Stir until smooth. Follow steps 2 and 3 above.

Activity #16 - Making Sedimentary Rock

Sedimentary rock is made of layers of differing sediments cemented together by chemical reactions. In the next three activities, you will demonstrate this layering process.

Materials:
Sand
Pebbles
Glue
Disposable cup; Clear would be best so you can see the layers of rock.

Procedure:

Alternate THIN layers of sand, glue, and pebbles until the cup is about one third full. Wait for rocks to dry. Let dry outside, if possible to speed up the process. Depending on the time of year, this could take up to a week. When you think it is dry, cut the cup off the rock. If it is not dry, let it continue to dry. What do you see? Why is this a sedimentary rock? Record your observations.

Sedimentary breccia

Activity #17 - Making Sedimentary Sandwiches

Materials:
Shale (from your rock kit)
2 pieces of bread (two different kinds/colors if you have them)
Favorite sandwich fillings using as many different items as possible
Condiments

Procedure:

First look at the sample of shale (sedimentary rock) with your child. Ask what the rock looks like; if they don't mention layers or stripes, add that into your description of the rock. Some sedimentary rock is composed of many layers pushed together and cemented over time. Today they are going to make a layered sandwich for lunch.

1. Present all the sandwich ingredients to your child and let them know that they are going to make their own lunch.

2. The ingredients can go in any order just as long as one piece of bread is on the bottom and one is on the top.

3. Let your child construct their own sandwich. Be sure that it has some sort of condiment (mayonnaise, mustard, peanut butter, jelly, etc.) or "cementing" agent. You may have to apply this if your child is unable to spread it.

4. If the sandwich has gotten tall you get the extra fun of letting your child "apply pressure" by squishing it down.

5. Make sure to look at the layers before getting down to the business of eating.

6. Make a second sandwich (not to eat). Set it aside to dry out.

Ask questions like the following -Did your sandwich look at all like the rock we looked at? Did you put all your layers on at once or did they go on one at a time? (Some sedimentary rock is formed a layer at a time, like shale.) When you bite through the sandwich does it all look the same or are there layers on the inside too? Try bending the layers while the sandwich is still moist. Set the second sandwich aside to completely dry. This may take several days. Try bending it when it is dry. What happens? Try this with your shale sample. Record your observations.

The Painted Desert, sedimentary layers, Arizona

Activity #18 - Crayon Rocks

Let's revisit the types of rocks again (sedimentary, metamorphic and igneous (plutonic and volcanic), and let's try making them from crayons.

Materials Needed:
Crayons, several of each in four different colors
Crayon or pencil sharpener or plastic knife
4 containers for holding the crayon shavings
3-6"x6" pieces of aluminum foil
Popsicle stick or other disposable stirrer
Mug
Boiling water

Procedure:

Unwrap the crayons and use the sharpener to create shavings. (You can also try to do this using something like a plastic knife.) Be sure to keep the shavings separate from each other at this point.

A. Sedimentary Rocks
Sedimentary rocks are formed from sediments (tiny rock particles that were created by weathering or erosion) that were layered compressed and cemented. To replicate this with the crayon shavings, take one of the aluminum foil squares and have your child sprinkle each of the colors of shavings into the middle of the square, one at a time so they will form the layers.

Fold the aluminum foil up tightly around the shavings and then compress it. To do this, they can press on it with their hands, step on it, place it in a clamp or use your creativity to think of other ways.

This does take a while to get the crayon pieces to stick together; a little body heat "helps" the process along. Carefully unfold the foil and remove the sedimentary rock with care as this is the most brittle of the rocks that you are making.

B. Metamorphic Rocks

Metamorphic rocks are formed when existing rocks are exposed to heat and/or pressure. To mimic this, take another square of aluminum foil and pile all four colors of shavings in the center. Fold up the sides of the aluminum foil to make a boat. Pour boiling water into a mug (adult job!) and float the boat in the hot water for 15 to 20 seconds, just until the shavings have started to melt.

Quickly remove the boat and fold the foil in half so that the shavings are compressed a bit. Let it cool and solidify and then open the foil and remove the "metamorphic rock."

C. Volcanic Rocks

Igneous rocks are formed when magma (molten rock) cools and solidifies. Now, we certainly know this to be true of volcanic rocks, but no one has seen plutonic rocks like granite form. So, geologists are guessing.

To make an igneous crayon rock, repeat the steps for making the metamorphic rock, except leave the aluminum foil boat floating on the hot water for a minute or more until all the crayon sediments have melted. (This may take more or less time depending on the brand of crayons that you have used.) Then take the popsicle stick and stir the shavings until they are all mixed together. Remove the foil from the water and let the crayon cool and solidify.

Record either the process of making each type of rock or draw a picture of what they looked like at the end (or both).

Activity #19 - Collecting Rocks

Materials:
Egg carton(s)
Paint, crayons or other art supplies to decorate the box

Procedure:

Paint or otherwise decorate your egg carton. This will be the beginning of organizing the rocks you collect. After your box is done, you can begin to transfer your treasures to the carton. You could have one box for each type of rock (igneous, metamorphic, sedimentary). Or you could classify them according to light and dark colors. You could even categorize them according to whether they are coarse grain or fine grain. If you know what kind of rocks they are, go ahead and label them. Write on a small piece of paper what the rock is, and put it beneath the rock in the carton. Be sure to include where you found it.

Volcanic basalt scoria

Materials list:

Activity notebook
Small bowl
Epsom salt
Sugar
Food coloring
Tape or pencil (any small stick)
Piece of twine
(Optional) weight of some sort - paper clip, washer, etc.
Vinegar or lemon juice
Magnifying glass
Two old towels
Four boxes of cereal, including Total Whole Grain
Very strong magnet
Rolling pin or blender
Gallon zip lock bags
Hot water
Sharpie marker
Hammer
Balloon
Plastic container or aluminum pie plate
Graham crackers
Chocolate bar
Marshmallows
Gummy bears/worms or similar candy
Paper plates
Large flat bottomed pan or bowl
Cooking spray
Baking chips - (white chocolate, peanut butter, butterscotch, chocolate)
Microwave Safe Bowl
Wax paper

Some type of mold to cool the chocolate in (can be a bowl, muffin pan, etc.
3 T. butter
1 package (10 oz.) marshmallows
6 C. Rice Krispies or similar cereal
Various candies or cake decorating materials
Glue
Clear disposable cup
Bread (two different kinds/colors if you have them)
Favorite sandwich fillings using as many different items as possible
Condiments
Crayons, several of each in four different colors crayon
Pencil sharpener or plastic knife
4 small containers
Aluminum foil
Popsicle stick or other disposable stirrer
Mug
Honey
Corn syrup
Molasses
Milk
Oil
Dish soap
Pancake syrup
Tray
A glass jar (holds 1-2 C.)
Small pitcher
A few small items, like marbles, paper clip, eraser, piece of rock, sand, etc.
Egg carton(s)
Art supplies, paints, crayons, etc.
Chunky cookie with variety of nuts, chips, and/or fruit pieces
Modeling clay in 4 colors (to simulate the earth's layers)

Toothpicks

Rocks and minerals:

- Quartz
- Amphibole
- Pyroxene
- Galena
- Calcium Feldspar
- Biotite Mica
- Magnetite - Iron
- Olivine
- Muscovite Mica
- Potassium Feldspar
- Sodium Feldspar
- Jasper
- Calcite
- Gabbro
- Granite
- Granodiorite
- Pegmatite
- Basalt
- Rhyolite
- Obsidian
- Pumice
- Scoria
- Gneiss
- Phyllite
- Quartzite
- Slate
- Shale
- Sandstone
- Limestone (3)

Patrick Nurre has been a rock hound since childhood and has an extensive rock, mineral and fossil collection, having collected from all over the United States. In 2005, he started

Northwest Treasures, which is devoted to designing geology kits for schools, and he has written ten textbooks for the study of geology. He conducts numerous field trips each year in Washington State to such places as the Olympic Peninsula, Mt. Rainier, Mt. St. Helens, the Channeled Scablands, Mt. Baker and Whidbey Island. In addition, he also gives field trips to the volcano loop of Oregon and California, Mt. Hood volcanic area (Oregon), the eastern badlands of Montana and Yellowstone National Park. In 2012, he opened the Geology Learning Center in Mountlake Terrace, WA. Patrick is a popular speaker at homeschool conventions, schools, and churches. He currently co-pastors a church in the Seattle, Washington area. Patrick and his wife, Vicki, have three children and one grandchild, and live in Bothell, Washington.

If you would like to contact Patrick about speaking or field trips:
northwestexpedition@msn.com
For a list of speaking topics: NorthwestRockAndFossil.com

42906939R00032

Made in the USA
Charleston, SC
08 June 2015